Robotic Fish iSplash-MICRO:
A 50mm Robotic Fish Generating the Maximum Velocity of Real Fish

By
Richard James Clapham PhD

iSplash Robotics

Published by *iSplash* Robotics
www.isplash-robotics.com
r.j.c@ieee.org

Copyright © 2016
Published by *iSplash* Robotics 2016
Illustration by Richard James Clapham PhD

All rights reserved. No part of this publication may be reproduced, stored in a retrieval system or transmitted in any form or by any means, electronic, mechanical, photocopying or otherwise without the prior permission of *iSplash* Robotics.

ISBN-13: 978-1537271064
ISBN-10: 1537271067

Thank you for your purchase.

The Worlds Fastest Fish

Abstract— This paper presents a millimeter scale robotic fish, namely iSplash-MICRO, able to accurately generate the posterior undulatory pattern of the carangiform swimming mode, at intensively high frequencies. Furthermore an investigation into anterior stabilization was made in an attempt to reduce the large kinematic errors and optimize forces around the center of mass. Applying large scale dorsal and pelvic fins relative to body size enabled predictable optimization of the anterior and posterior displacements. During the field trials, the small fish with a length of 50mm has generated an equivalent average maximum velocity to real fish, measured in body lengths/ second (BL/s), greatly improving previous man-made systems, achieving a consistent free swimming speed of 10.4BL/s (0.52m/s) at 19Hz with a low energy consumption of 0.8 Watts.

Keywords: Micro Robotics • Marine Robotics • High speed robotic fish • Carangiform swimming • Full-Body length • Swimming speed • Maximum velocity.

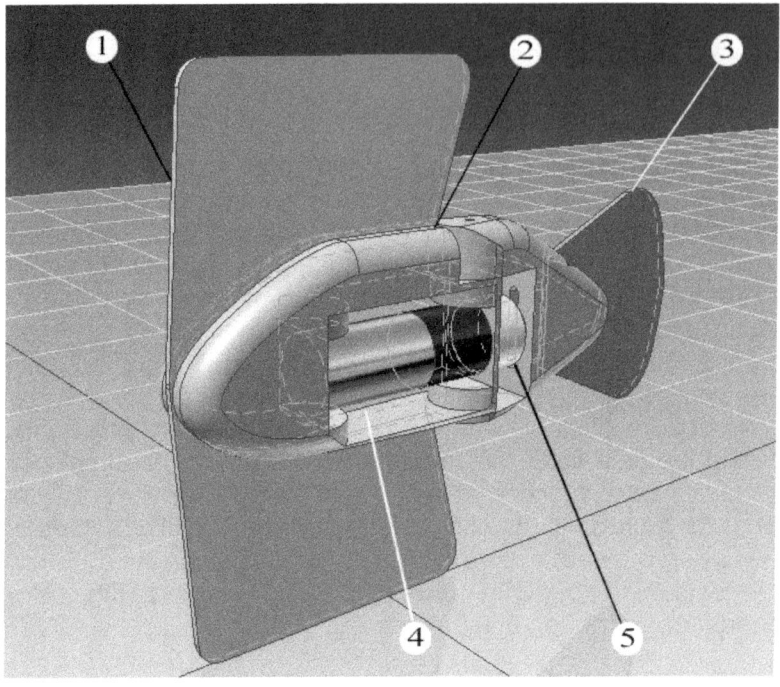

Figure 1. *iSplash*-MICRO, 50mm in body length: 1-Large scale anterior fins; 2-Modular mid-body chassis; 3-Compliant caudal fin and peduncle; 4-Power transmission system; 5-Driven tail plate

I. INTRODUCTION

During marine exploration a miniaturized underwater vehicle (UV) can benefit operations by allowing greater mobility as the turning diameter and body geometry is small. Current rotary propeller driven vehicles operating during low speed locomotion have particularly high cost of transport [1] requiring large a power supply, increasing body size. In addition the drive method of rigid hull UVs create large propulsive wakes causing increased environmental noise. In contrast, this is where fish excel, generating large transient forces efficiently and smoothly by coordinating their body motion in harmony with the surrounding fluid [2],[3]. Hence, the miniaturization of bio-robotic propulsion can provide greater mobility, unobtrusive navigation, reduced noise to the environment and a lower cost of transport.

Although smaller fish are measured with an overall slower speed in relation to size, they are able to generate higher tail oscillatory frequencies and therefore greater speed, measured in body lengths/ second (BL/s), a great benefit to overcome position destabilization flows. Bainbridge has measured live fish with average maximum velocity of 10BL/s, an exceptional example of a small fish (Cyprinus carpio), 135mm in body length, achieved a maximum velocity of 12.6Bl/s (i.e. 1.7m/s) and a stride rate of 0.7 [4][5].

Micro scale design requires developing the propulsive motion in relation to the developed body size and desired surrounding environment. The required propulsion method can be calculated using the Reynolds number (Re), defined as Re=UL/v, where U denotes the speed, L denotes the body length and v is the viscosity of water. Re relates to the ratio of inertia to viscous forces. If the Re is large, the viscous forces are negligible (Re 10^3-10^7) [6]. Viscous force is dominant when the body length is approximately <1mm. The two methods of propulsion are categorized as: (i) Inertia Force Propulsion, generating locomotion by creating a reaction force (FR) against the mass of the water. (ii) Resistance Force Propulsion, adopting a kinematic motion to generate locomotion from the viscosity of the water.

Developing robotic fish at any body-length has proven a great challenge. Although most work has focused on hydrodynamic mechanisms, performance is still low. In particular: (i) Accurately replicating the linear swimming motion has proven to be difficult and free swimming robotic fish have significant kinematic parameter errors. (ii) Actuator selection and mechanical transfer into productive propulsion are limited by force, frequencies and mechanical losses.

The complexity of developing a mechanical structure at the micro scale is increased due to material and hardware constraints. Some examples of novel micro builds and their maximum speeds are Ye's IPMC actuated fish, 98mm's in body length which achieved a maximum velocity of 0.24BL/s (24mm/s) [7], Guo's ICPF actuated prototype, 45mm's, achieving a velocity of 0.11BL/s (5.21mm/s) [8] and Wang's SMA manta ray, 243mm's, achieving a velocity of 0.23BL/s (57mm/s) [9]. Some examples of novel design approaches at larger scales are Barrett's Robotuna, which achieved a velocity of 0.65BL/s (0.7m/s) [10], Yu's discrete assembly achieving a velocity of 0.8BL/s (0.32m/s) [11], Liu's, G9 achieving a velocity of 1.02BL/s (0.5m/s) [12] and Valdivia y Alvarado's compliant structure achieving a velocity of 1.1BL/s (0.32m/s) [13]. Previously the low speeds of robotic swimmers were unpractical for operation, peaking at speeds of 1Bl/s.

iSplash-I [14], i.e. a carangiform swimmer with a body length of 250mm, introduced a high performance swimming motion. The novel mechanical drive system was devised to operate in two swimming patterns. A thorough comparison between the traditional posterior confined undulatory swimming pattern and an introduced full body length swimming pattern was made. The proposed pattern coordinated anterior, mid-body and posterior displacements, reducing the large kinematic errors seen in free swimming robotic fish. In particular it achieved anterior stabilization by significantly reducing recoil and optimizing the lateral (FL) and thrust forces (FT) around the center of mass, assumed to initiate the starting moment of added mass upstream. From the comparison, the proposed swimming motion significantly outperformed the traditional approach, achieving a maximum velocity of 3.4BL/s (0.88m/s) and consistently achieved a velocity of 2.8BL/s (0.70m/s) at 6.6Hz. Notably, the mechanical drive system of iSplash-I operating in the simplified body motion of posterior confined undulations also measured a high performance of 2.2BL/s (0.55m/s) at 6.1Hz.

Throughout the field trials the prototype showed no peak or decline in velocity as frequency was raised in both swimming modes. Mimicking the swimming variables of real fish, as measured in the observational studies of Bainbridge, indicating swimming above 5Hz has no variation in kinematics, and only frequency is changed to increase swimming speed, providing a key swimming parameter. Therefore applying higher frequencies to either swimming motion of iSplash-I may continue to increase velocity, and in consideration that fish employ an identical swimming motion of inertia force propulsion at the millimeter scale, we proposed iSplash-MICRO.

The remainder of the paper is organized as follows: Section II details the applied swimming pattern and introduces a novel destabilization solution. Section III describes a new construction method for micro builds. Section IV describes the field trials undertaken and the experimental results obtained. Concluding remarks and future work are given in Section V.

A. Research Objectives.

The research project aimed to achieve the fastest speeds of live fish and proposed five main objectives:
(i) Develop a structural build only 50mm in length, able of accurately replicating the kinematic parameters of the posterior confined undulatory swimming pattern of iSplash-I;
(ii) Significantly raise driven frequency in comparison to the first generation, to be capable of intensive tail oscillations of up to 19Hz by fabricating a robust naturally buoyant structure, a complex challenge at such a small scale;

(iii)   Allow for a high efficiency mechanical energy transfer by engineering a drive system that takes hardware and material parameters into account;

(iv)   Devise a novel solution to improve the predicted excessive destabilization in yaw, by optimizing the FL and FT around the center of mass;

(v)   Realize a mechanism capable of consistent steady state swimming, measuring its achievements in terms of speed, kinematic accuracy and energy consumption over a range of frequencies from 5-19Hz.

II. DESIGN METHODOLOGY

A. Posterior Undulatory Swimming Motion

The selected carangiform swimmer, Cyprinus carpio (common carp), which applies the method of body and/or caudal fin (BCF) propulsion has been chosen for replication due to its exceptionally high locomotive performance [4],[5].

Inertia Force Propulsion of the Carangiform swimming mode is associated with the method of added mass [15]. Added mass is initiated as each individual segment of the undulatory wave passes backwards during the body wave cycle creating a force F_R against the surrounding fluid and an opposing force against the body, generating forward motion of the entire body. The F_R is decomposed in to F_T and F_L, which must be optimized for efficient propulsion. The added mass is the product of the water accelerated and the momentum of water accelerated by the propulsive segments. Fish can achieve a low cost of transport by generating the method of added mass efficiently. Hence, there is great potential to improve biological inspired UV's by accurately mimicking this propulsive method.

The kinematic pattern of the carangiform inertia force propulsion method is represented in the form of a traveling wave and can be identified by its propulsive wave length and amplitude envelope. Traditionally robotic swimmers adopt a method which typically concentrates the wave motion to <1/2 of the body length, initiated at the center of mass, smoothly increasing in amplitude along the body length towards the tail tip [2]. The observed posterior amplitude of live fish is 0.1 of the body length, measured from the midline to the furthest lateral tail excursion. The location of the pivot point is optimum at 0.15-0.25 [4]. The commonly adopted swimming kinematics were proposed in [10] adapted in [12]. The posterior undulatory swimming motion applied to iSplash-MICRO is of the form:

$$y_{body}(x,t) = (c_1 x + c_2 x^2)\sin(kx + \omega t) - c_1 x \sin(\omega t) \quad (2)$$

where y_{body} is the transverse displacement of the body; x is the displacement along the main axis starting from the nose of the robotic fish; k =2π/λ is the wave number; λ is the body wave length; ω = 2πf is the body wave frequency; c_1 is the linear wave amplitude envelope and c_2 is the quadratic wave. The desired wave form motion consists of one positive phase and one negative phase throughout the complete cycle [4]. The parameters P = {c_1,c_2,k,ω} can be adjusted to achieve the required posterior swimming pattern, providing an engineering reference. This operational mode will be described as Mode 1 and is illustrated in Fig. 6.

B. Anterior Stabilization Approach

It was detailed by Lighthill [2], that several morphological adaptations of carangiform fish reduce the severe yaw destabilization that is seen in free swimming robotic fish, without affecting propulsive efficiency by increasing drag: (i) Vertical compression along the full body length. The measured maximum thickness is 0.2 of the body length and reduces in size towards either extremity [4]. This form must be applied to reduce roll and yaw destabilization and also forward resistance. This structure is realized when engineering a morphological approximation of a carangiform swimmer; (ii) Reduced depth of body at the peduncle, most defined by the thunniform swimming mode [10]. This structural adaptation is complex to fabricate at any scale due to material limitations, particularly at the micro scale as the posterior mechanism must be narrow, structurally robust and mimic the smooth curves of the kinematic undulations; (iii) Accurate weight distribution. Imprecise configurations generate large destabilization within the horizontal plane, as a consequence of the F_R not being optimized, typically seen on multi-link servo assemblies [11],[12]; (iv) Increased depth of body towards the anterior, a distinctive adaptation of the carangiform morphological parameters. As previously described, anterior destabilization is challenging to control [14]. Passive rigid anterior mechanisms recoil around the center of mass. Current free swimming robotic fish can be seen with excessive head swing, similar in magnitude to the posterior, greatly increasing drag [13]. As a micro scale structure limits the ability to apply the coordinated full-body motion [15] due to the complexity of the mechanical linkage, we propose a novel approach to significantly increase the anterior depth of body by applying large scale dorsal and pelvic fins, relative to body size.

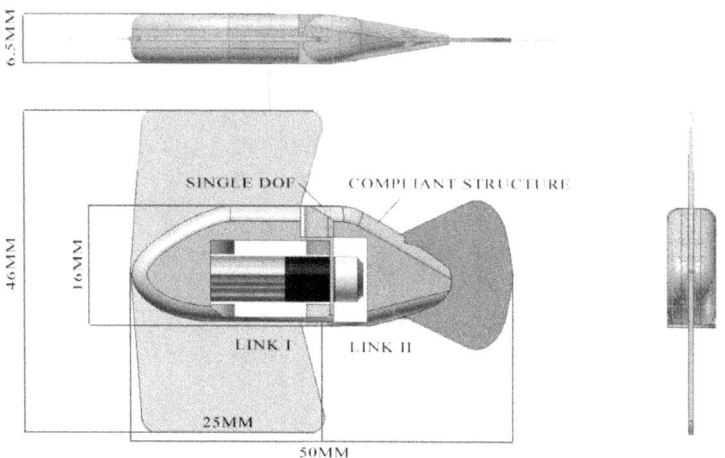

Figure 2. 1-Plan; 2-Side; 3-Front view.

The anterior destabilization solution of anterior and mid-body fins provides the required surface area without creating excessive resistance to forward motion as the fins are vertically compressed. Flow visualization techniques from published biological studies [16][17] and mechanisms such as "undulating pump" and "vortex peg" have measured a fluid-body interaction, that may contribute to propulsion, is generated upstream to the posterior section. Therefore, it is predicted that reducing the anterior amplitude to the observed optimal value of a common carp at approximately 0.04 of the body length will initiate the starting moment of added mass upstream and optimize the F_L and F_T forces around the center of mass, increasing the overall magnitude of thrust contributing to increased forward velocity. This operational mode will be referred to as Mode 2, illustrated in Figs. 2 and 7.

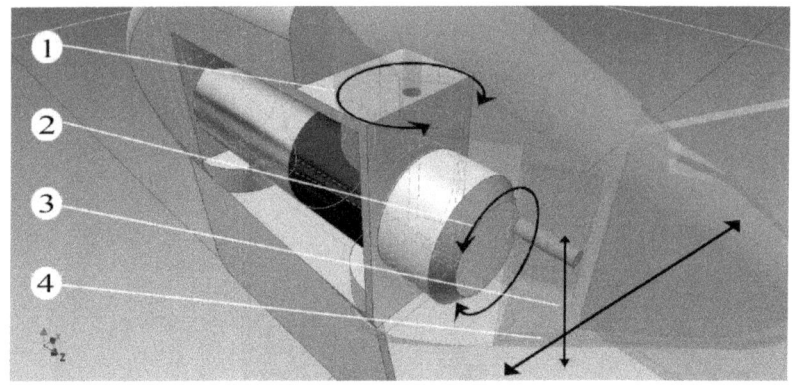

Figure 3. Power transmission system: 1-Tail connecting pivot; 2-Crankshaft; 3-Transistion plate; 4-Link II coupled to caudal fin.

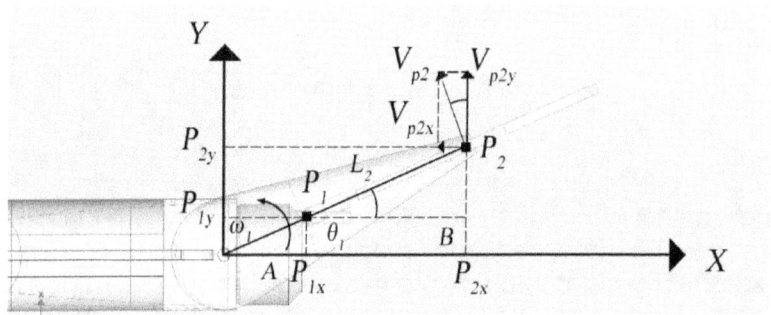

Figure 4. Schematic drawing of the tail offset drive crank and linkage;

III. CONSTRUCTION METHOD

A. Mechanical Design.

A mechanical drive system able to accurately mimic the displacements of the travelling wave sets a great challenge. This is significantly increased at the millimeter scale. The drive system must be simple, compact, lightweight and robust to provide high frequencies under large forces for high-speed performance. Various novel approaches using the advances in hardware to achieve micro scale actuation have been attempted [18],[19],[20]. Smart materials have mainly been adopted to reduce body size but have shown limited frequencies, response time, actuation and force, therefore measuring low linear velocities [7],[8],[9],[21].

We set an aim for the prototypes build length of 50mm. Live carangiform fish of this body size employ identical swimming motions to larger fish of 250mm to generate linear locomotion, by employing the inertia force propulsion method of added mass. Therefore, we aimed to devise a simplified powertrain, capable of replicating the posterior confined undulatory swimming motion of iSplash-I without loss of accuracy.

The developed arrangement illustrated in Fig. 2 employs a single electric motor and a single degree of freedom along the axial length. The two rigid discrete links are coupled with a stiffness varying posterior, beginning at link II and continuing to the tail tip. The discrete assembly method can be defined as a series of links or N links. Applying more links reduces midline curve alignment errors and increases mechanical complexity. Details of the fully discretized body wave fitting method are given in [11], [12]. To provide the undulatory motion, a compliant caudal fin is coupled to link II which can be removed. This method allowed the material stiffness of the caudal fin to be adjusted experimentally, to provide the targeted curves during free swimming. The presented design approach achieved complexity of motion with a significant reduction in structural parts, allowing a build only 50mm in length to be realized.

Figure 5. *iSplash*-MICRO with large dorsal and pelvic fins.

In addition, a modular build was devised allowing each part to be easily changed and modified during fabrication, and for all Modes of operation to be applied to the same prototype by adjusting the configuration. Mode 3 can be put in operation by adjusting the single offset crank, driving link II shown in Fig. 3. Mode 2 can be applied by removing the outer structure of link I and replacing with the required form.

B.   Power Transmission System.

A single crankshaft attached to the output shaft of the primary actuator directly drives link II, by transmitting continuous rotary power to linear oscillations, illustrated in Fig. 3 and 4. The devised powertrain increases power distribution to the posterior and required high-precision of the chassis, link structures and crankshaft to avoid deadlock, reduce friction and significantly improve energy transfer, which can be lost in many stages of the mechanical drive.

The driven link amplitude is determined by the offset crank, L_2 represents posterior link of the discrete structure. The maximum amplitude of the link length L_2 at point P_2 is determined by the predetermined maximum crank offset P_1. The coordinates of P_1 (P_{1x}, P_{1y}) and P_2 (P_{2x}, P_{2y}) can be derived by:

$$\begin{cases} P_{1x} = A\cos\theta_1 \\ P_{1y} = A\sin\theta_1 \end{cases} \begin{cases} P_{2x} = P_{1x} + B\sin\theta_1 \\ P_{2y} = P_{1y} + B\sin\theta_1 \end{cases} \quad (3)$$

The length of L_2 can be derived by $L_2^2 = P_{2x}^2 + P_{2y}^2$. Assume that ω_1 is the angular velocity of the link L_2, and the velocity vector V_{P2} is perpendicular to L_2. We have:

$$\begin{cases} V_{p2x} = -\omega_1 L_2 \sin\theta_1 \\ V_{p2y} = \omega_1 L_2 \cos\theta_1 \end{cases} \quad (4)$$

where V_{p2x} and V_{p2y} are the decomposed vectors of the velocity vector $V_{P2} = \omega_1 L_2$.

C. Fabrication.

iSplash-MICRO in operational Mode 2 with a total mass of 3.35g (2.45g in Mode 1) is illustrated in Fig. 5, with the physical specifications given in Table I. The build was first digital modelled taking hardware constraints and material properties into account, so that the kinematic and geometric parameters were not affected, reducing the resistance during forward locomotion. The inner structural frames and bulkhead were formed from the material polypropylene, chosen to provide the required density (lower than water) and the structural strength for high frequency actuation. All structural parts were hand cut, fitted and assembled. This devised method produced a structurally robust prototype allowing for consistency of operation at intensively high frequencies, generating large forces relative to body size, which are applied to the water and reactively, the opposing force exerted on to the vehicle [15].

For fast locomotive performance, a high power density is required to attain high frequency actuation. Although complexity of developing the mechanical drive system is increased, a single electrical motor with continuous actuation was deployed and positioned in the optimum location [2], [4], in contrast to other construction methods which limit accuracy of weight and volume arrangement and force production [11],[12],[13]. The power-to-weight ratio was greatly increased by achieving a configuration in which the electric motor is 50% of the total mass in operational Mode 1 and 37% in Modes 2 and 3. To counteract the large weight of the primary actuator relative to body size, the build volume was increased vertically, to realize natural buoyancy.

TABLE I. PHYSICAL PARAMETERS OF *ISPLASH*-MICRO

Parameter	Specific Value
Maximum Velocity: BL/s (m/s)	10.4 (0.52)
Body Size: mm (LxWxH)	50 x 7 x 16 (Without Fins)
Body Mass: g	2.45 (Without Fins)
Anterior Fin Size: mm (LxWxH)	26 x 0.8 x 48
Anterior Fin Mass: g	0.9
Maximum Frequency: Hz	19
Actuator:	Single electric motor
Actuator Mass: g	1.23
Power Supply:	4V LiPo external battery supply
Materials:	Balsa wood, polypropylene,
Swimming Modes:	Linear locomotion
Tail Material:	Polypropylene
Thickness of Caudal Fin : mm	0.8
Caudal Fin Aspect Ratio: AR	1.5

For operational Mode 2, large fins, 48mm in height with vertical compression were devised, a similar morphological trait as real fish fins. By forming large dorsal and pelvic fins made of polypropylene the buoyancy was greatly increased, this was counteracted by applying weight to the lower chassis of link l. During free swimming stability could not be maintained as frequencies were raised. Although the anterior fins were devised as an initial solution to reduce destabilization within the horizontal plane, the configuration of material properties significantly increased stabilization in roll. This development enabled the prototype to achieve open loop stability, as the relative position of buoyancy is higher than the center of mass. Therefore the surrounding fluid counterbalances the gravitational weight [22].

IV. Experimental Procedure and Results

A. Field Trials.

Experiments were conducted within a 1000mm long x 500mm wide x 250mm deep test tank. The prototype had sufficient space to move without disturbances from side boundaries and the free surface, able to maintain swimming at mid-height of the test tank. A series of experiments were undertaken in order to verify the locomotive performance of all Modes in terms of speed, energy consumption and kinematic observation at frequencies between 5-19Hz. The test results of all Modes are summarized in Table II. Steady free swimming over a fixed distance of 450mm was used to measure speed (shown in Fig. 9), once the prototype had reached its maximum velocity at each tail oscillation frequency. Measurements were averaged over many cycles once the prototype was able to achieve consistent operation.

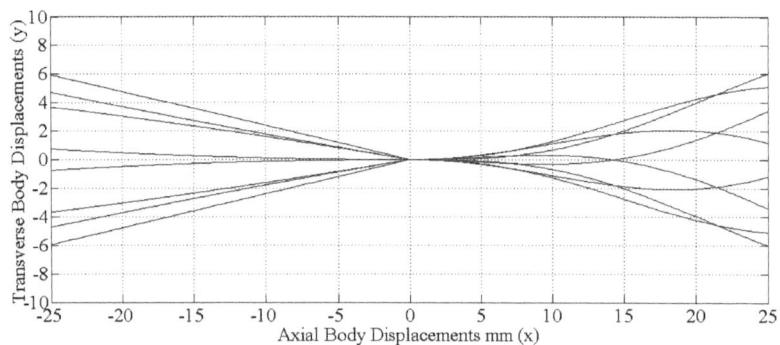

Figure 6. Mode 1: Kinematic data of midline during free swimming. 6mm tail amplitude and 6mm head amplitude (0.12 of the body length).

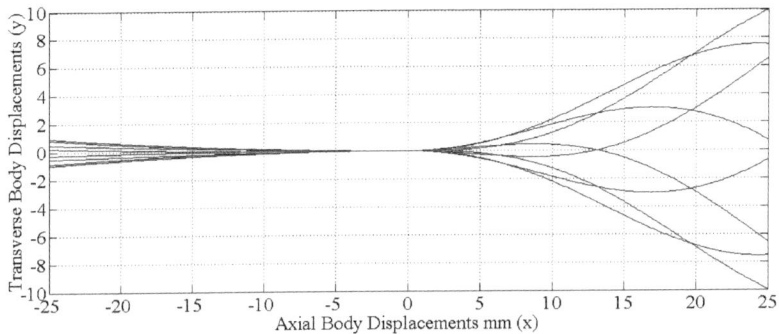

Figure 7. Mode 2: Appling large scale dorsal and pelvic fins relative to body size to reduce yaw destabilization, generating 1mm head amplitude and 10mm tail amplitude (0.20 of the body length).

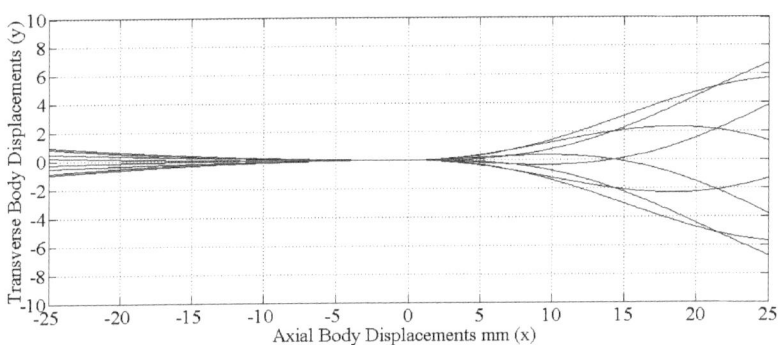

Figure 8. Mode 3: Operating with a reduced tail amplitude of 7mm (0.14 of the body length).

B. Swimming Pattern Observation.

Illustrated in Figs. 6, 7 and 8 are the midline kinematic parameters of the full body length, taken at every 0.006s to provide the anterior and posterior wave forms of Modes 1-3 during swimming for comparison. We aimed to achieve and analyze the kinematic parameters of live fish and *iSplash*-I. Good matching with live fish kinematic data is a complex task, as the wave forms of free swimming robotic fish have shown extensive head amplitude errors and inaccurate displacement parameters over the full body length.

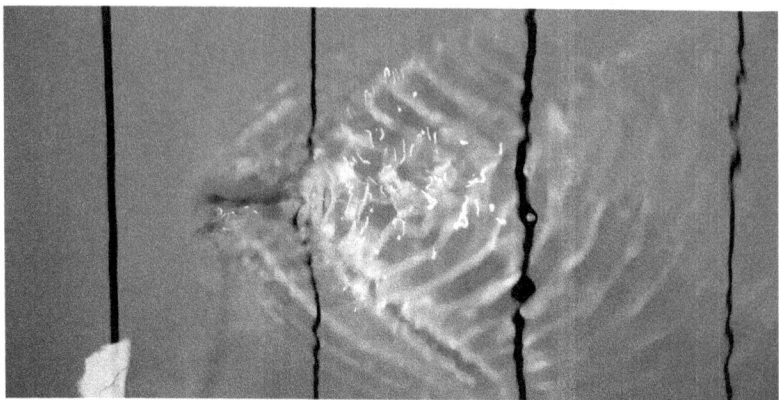

Figure 9. *iSplash*-MICRO during experimental testing of maximum velocity at 19Hz. Measured over a fixed distance of 450mm.

When comparing Modes 1 and 2, Mode 2 reduced the head amplitude from 0.12 (6mm) of the body length in Mode 1 to 0.02 (1mm), generating a smaller value than the common carp at 0.04 [4]. The observed tail amplitude of the common carp is 0.1 and *iSplash*-I is 0.17. Mode 1 with anterior destabilizing, lost posterior amplitude displacements and measured 0.12 (6mm), in comparison Mode 2 was able to generate a large value of 0.20 (10mm). Analyzing the midline sequence of the Mode 1 illustrated in Fig. 6, we can notice the excessive head swing amplitude equivalent to the posterior value as predicted, based on previous work in [15] and current robotic fish [12][13]. The stabilizing solution of Mode 2 has significantly improved matching errors, greatly reducing the recoil from the concentrated posterior thrust and in turn creating accurate posterior displacements. Fig. 8 shows that Mode 3 has a reduced posterior amplitude of 0.14 (7mm), closer to the observed measurement of live fish and an equivalent anterior displacement to Mode 2. We were able to calculate the tail amplitude of Mode 3 prior to the experimental testing using (3), as the devised anterior fins of the free swimming robotic fish have enabled a predictable kinematic optimization. In addition, it was found that Mode 1 had an erratic forward heading during test runs, affected by the tethered cables. In contrast, Mode 2 provides a very directional forward heading, not deterring once in motion.

A principal aim of the research was achieving accurate posterior kinematic matching to *iSplash*-I, with a significantly simplified assembly. The tracked midline shows that the developed prototype generates a precise travelling wave ½ the body length and has achieved a smooth transition phase from body to tail tip, due to the compliant structure. The formed caudal fin provided higher speeds in initial testing with a low aspect ratio, future optimization is planned. AR is calculated using: $AR = b2/Sc$ where b squared is the fin span and Sc is the projected fin area. AR in this case was 1.6.

COMPARISON OF TEST RESULTS BETWEEN MODES 1, 2 & 3

Parameters	Mode 1	Mode 2	Mode 3
Reynold Number: Re (10^4)	2.6	2.6	1.6
Shrouhal Number: St	0.22	0.36	0.41
Swimming Number: Sw	0.54	0.54	0.33
Maxium Velocity: BL/s	10.4	10.4	6.4
Maxium Velocity: m/s	0.52	0.52	0.32
Frequency: Hz	19	19	19
Max Power Comsumption Air: W	0.68	0.68	0.68
Max Power Comsumption Water: W	0.8	0.8	0.8
Head Swing Amplitude: mm	6	1	1
Tail Swing Amplitude: mm	6	10	7
Test Run Distance: mm	450	450	450

C. Experimental Results.

The average energy consumption in relationship to driven frequency is shown in Fig. 10. A comparison measuring the cost of transport in water and actuation air was made. We can notice that all Modes actuating in water resulted in a slight increase in energy consumption, from 0.6W in air to 0.8W, at the maximum frequency of 19Hz. The very low cost of transport measured indicates the replicated swimming motion of [15] and the novel mechanical drive system is energy efficient. We can calculate that the next generation will be capable of carrying an onboard power supply.

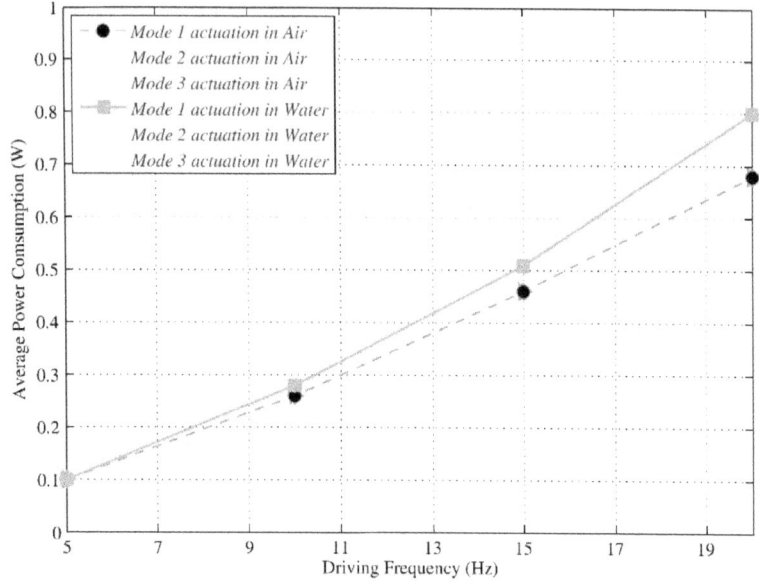

Figure 10. Comparison of average electrical power consumption over driven frequency of all Modes actuating in air and water.

The relationship between velocity (speed divided by body length) and driven frequency is shown in Fig. 11. The corresponding values of all Modes during consistent swimming were measured and compared to current robotic fish. Modes 1 and 2 consistently achieved a maximum velocity of 10.4Bl/s (0.52m/s) at 19Hz. Mode 3 had a decreased maximum velocity of 6.4BL/s (0.32m/s) at 19Hz. Although the tail amplitude of Mode 3 was closer to the value of the common carp, performance was lower, reaffirming data from the first generation [15]. In addition, no energy consumption was saved from the tail amplitude reduction. Hence, we estimate the next generation may have enough power to further increase of tail amplitude beyond the value of Mode 2 (the furthest lateral excursion that could be applied to this structure). iSplash-MICRO has significantly increased performance in comparison with current robotic fish which typically peak at approximately 1BL/s and has equivalent speeds of the fastest real fish, measured with an average maximum velocity of 10BL/s [5].

Identical to the previous generation as frequencies were raised, velocity increased in all Modes. From this we can assume that raising this key swimming parameter of the developed mechanical drive system beyond 19Hz will continue to increase performance further.

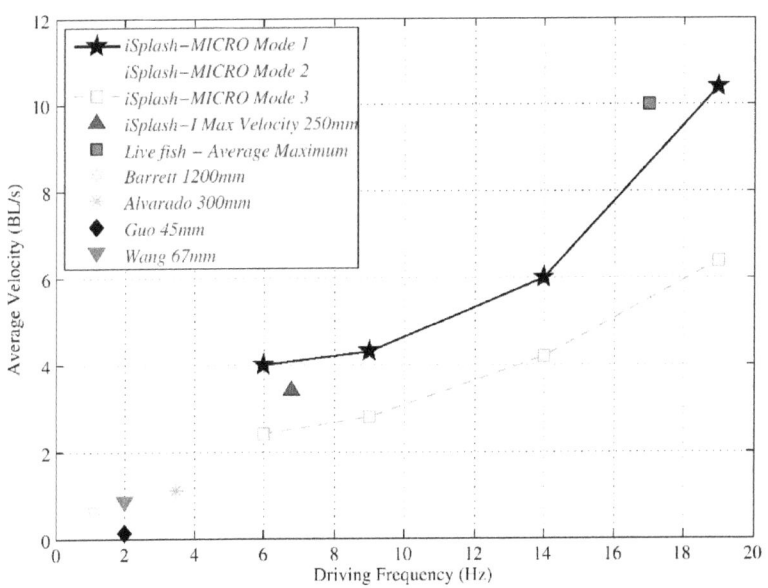

Figure 11. Comparison of average velocities achieved by all Modes, contrasted against current robotic fish and real fish.

Despite Mode 2 significantly improving displacement parameters along the full body length, no noticeable increase or decrease in velocity was found. This result was not expected as the coordinated full-body swimming motion of the first generation increased velocity by 27%. Differences in structure can be noted as previously mentioned in Section II-A, as a consequence the midline kinematics do not generate a smooth curve from head to tail. This may have stalled the fluid flow interaction at a point located around the center of mass. The location of the pivot point is inaccurate, due to structural constraints, the optimum range being 0.15-0.25 of the body length. All Modes were measured at 0.5.

A prominent parameter for analyzing BCF locomotive performance is the Strouhal number (St), defined as $St=fA/U$, where f denotes the frequency, A denotes the tail amplitude and U is the average forward velocity. The St is considered optimal within the range of $0.25 < St < 0.40$ [23]. Mode 1 has a St of 0.22 under the condition of $Re = 2.6 \times 10^4$, Mode 2 has a St of 0.36 under the condition of $Re = 2.6 \times 10^4$ and Mode 3 measured a $St = 0.41$ under a condition of $Re = 1.6 \times 10^4$. The St of Mode 2 is within the desired range with both Modes 1 and 3 approximately optimal. The prototypes Swimming number (Sw) (distance travelled per tail beat) of Modes 1 and 2 is highly efficient, measuring a Sw of 0.54, increasing performance over the previous build with a Sw of 0.42 and close to the particular efficient common carp with a Sw of 0.70 [4].

In addition, the prototype was actuated at high frequencies for long periods during construction and test runs, without need of rebuild or showing wear, indicating the robustness of the developed micro prototypes mechanical drive system

V. CONCLUSION AND FUTURE WORK

This paper describes the development of a small robotic fish prototype iSplash-MICRO, a man-made system able to generate the average maximum velocity of real fish. A greatly simplified structural built at the millimeter scale is able to accurately replicate the kinematic parameters of the posterior confined undulatory swimming pattern of iSplash-I [15]. Considering that the kinematic model over the full body length has large matching errors, a novel stabilization technique was deployed. The large scale dorsal and pelvic fins relative to body size optimize the FL and FT around the center of mass, generating accurate anterior amplitude and effectively stabilizing the platform in the horizontal and vertical planes. This enabled a predictable adjustment of the anterior and posterior kinematic amplitude parameters so that large posterior propulsive forces and amplitudes could be generated and accurate straight-line trajectories attained.

The developed prototype is compact, natural buoyant and robust, with a high power density and the ability to actuate at intensively high frequencies. Notably, the applied high frequencies increased velocity in all Modes without peak, decline or failure. We can therefore estimate that a further increase of frequency applied to this developed mechanical drive system may continue to increase its maximum velocity beyond the fastest live fish. iSplash-MICRO 50mm in body length achieved a consistent free swimming speed of 10.4BL/s (i.e. 0.52m/s) at a frequency of 19Hz with a stride rate (S_w) of 0.54 and a low energy consumption of 0.8W.

The experimental analysis has shown potential to improve the high performance swimming further, the following points are of significant interest: (i) The low energy consumed during propulsion indicates the next generation is capable of carrying a power supply; (ii) Posterior amplitudes of a magnitude greater than 0.2 of the body length may improve propulsion; (iii) The robust structure may generate higher tail oscillation frequencies. An initial aim of a 50% increase can be made, predicting an equivalent increase in performance; (iv) Optimize the tail shape and aspect ratio [22]; (v) The devised mechanical system is suitable for further reduction in scale as appropriate hardware becomes available.

ACKNOWLEDGMENTS

Our special thanks go to Richard Clapham senior for his technical assistance towards the project.

REFERENCES

[1] P. R. Bandyopadhyay, "Maneuvering hydrodynamics of fish and small underwater vehicles," Integr. Comparative Biol., vol. 42, no. 1, pp. 102–17, 2002.
[2] J. Lighthill, "Mathematical Biofluiddynamics", Society for Industrial and Applied Mathematics, Philadelphia, 1975.
[3] J. J. Videler, "Fish Swimming", Chapman and Hall, London, 1993.
[4] M. Nagai. "Thinking Fluid Dynamics with Dolphins," Ohmsha, LTD, Japan, 1999.
[5] R. Bainbridge, "The Speed of Swimming of Fish As Related To Size And To The Frequency and Amplitude of The Tail Beat", J Exp Biol 35:109–133, 1957.
[6] P.W. Webb, "Simple physical principles and vertebrate aquatic locomotion", Amer. Zool., vol. 28, pp.709 -725 1988.
[7] X. Ye, Y. Su, S. Guo and L. Wang, "Design and Realization of a Remote Control Centimeter-Scale Robotic Fish", Proceedings of the IEEE/ASME International Conference on Advanced Intelligent Mechatronics, pp. 25-30, 2008.
[8] S. Guo, T. Fukuda and K. Asaka. "Fish-like Underwater Microrobot with 3 DOF", In Proceedings of IEEE International Conference on Robotics and Automation, Washington, USA, pp. 738–743, May 2002.
[9] Z. Wang, Y. Wang, L Jian and G. Hang, "A micro biomimetic manta ray robot fish actuated by SMA", Proc. IEEE Int. Conf. on Robotics and Biomimetics (Guilin, February) pp 1809–13, 2009.
[10] D. S. Barrett, M. S. Triantafyllou, D. K. P. Yue, M. A. Grosenbaugh, and M. J. Wolfgang, "Drag reduction in fish-like locomotion," J. Fluid Mech., vol. 392, pp. 183–212, 1999.
[11] J. Yu, M. Tan, S. Wang and E. Chen. "Development of a biomimetic robotic fish and its control algorithm," IEEE Trans. Syst., Man Cybern. B, Cybern., 2004,34(4): 1798-1810
[12] J. Liu and H. Hu, "Biological Inspiration: From Carangiform fish to multi-Joint robotic fish," Journal of Bionic Engineering, vol. 7, pp. 35–48, 2010.
[13] P. Valdivia y Alvarado, and K. Youcef-Toumi, "Modeling and design methodology for an efficient underwater propulsion system", Proc. IASTED International conference on Robotics and Applications, Salzburg 2003.
[14] R. J. Clapham and H. Hu, "iSplash-I: High Performance Swimming Motion of a Carangiform Robotic Fish with Full-Body Coordination," Accepted for 2014 IEEE International Conference on Robotics and Automation, May 31 - June 7, 2014, Hong Kong, China.
[15] P. W. Webb, "Form and function in fish swimming," *Sci. Amer.*, vol. 251, pp. 58–68, 1984.
[16] M. W. Rosen, "Water flow about a swimming fish," China Lake, CA, US Naval Ordnance Test Station TP 2298, p. 96, 1959.
[17] M.J. Wolfgang, J.M. Anderson, M.A. Grosenbaugh, D.K. Yue and M.S. Triantafyllou, "Near-body flow dynamics in swimming fish," September 1, 1999, J Exp Biol 202, 2303-2327
[18] X. Deng and S. Avadhanula, "Biomimetic micro underwater vehicle with oscillating fin propulsion: System design and force measurements," Proc.

of IEEE International Conference on Robotics and Automation, pp.3312-3317, April, 2005.
[19] T. Fukuda, A. Kawamoto, F. Arai and H. Matsuura, "Mechanism and Swimming Experiment of Micro Mobile Robot in Water", Proc. IEEE Con. on Robotics and Automation, Vol.1, pp.814-819, San Diego, California, May 1994.
[20] E. Lee, "Design of a soft and autonomous biomimetic micro-robotic fish" In Proceedings of IEEE Industrial Electronics and Applications (ICIEA), pp. 240 – 247, June 2010.
[21] C. Rossi, W. Coral, J. Colorado and Barrientos, "A motor-less and gear-less bio-mimetic robotic fish design", Proc. IEEE Int. Conf. on Robotics and Automation, pp 3646–51, Shanghai, May 2011.
[22] G. V. Lauder and E. G. Drucker, "Morphology and Experimental Hydrodynamics of Fish Control Surfaces," IEEE J. Oceanic Eng., Vol. 29, Pp. 556–571, July 2004.
[23] G. S. Triantafyllou, M. S. Triantafyllou, and M. A. Grosenbauch, "Optimal thrust development in oscillating foils with application to fish propulsion," J. Fluids Struct., vol. 7, pp. 205–224, 1993.

iSplash Robotics

Published by *iSplash* Robotics UK
www.isplash-robotics.com
Copyright © 2016

Published by *iSplash* Robotics 2016
Illustration by Richard James Clapham PhD

All rights reserved. No part of this publication may be reproduced, stored in a retrieval system or transmitted in any form or by any means, electronic, mechanical, photocopying or otherwise without the prior permission of Natural Classics.

ISBN:

Thank you for your purchase.

The Worlds Fastest Fish

www.ingramcontent.com/pod-product-compliance
Lightning Source LLC
Chambersburg PA
CBHW070341190526
45169CB00005B/1989